NUMBERS

LEVEL 5

THE YOUNG SCHOLAR

MATHS WORKBOOK

AL & LES ASEERVATHAM

2007

THE YOUNG SCHOLAR

Contents

Level 4

Level Statement 2

1 **Number Concepts** 2
- Exercise 1.1 (Integers) 4
- Exercise 1.2 (Integer addition and subtraction) 7
- Exercise 1.3 (Integer multiplication) 10
- Exercise 1.4 (Square numbers and square roots) 16
- Exercise 1.5 (Rational and irrational numbers) 18
- Exercise 1.6 (Indices) 20
- Exercise 1.7 (Rates) 24
- Exercise 1.8 (Rates) 25
- Exercise 1.9 (Ratios) 28
- Exercise 1.10 (Ratios) 29
- Exercise 1.11 (Direct proportions) 32
- Exercise 1.12 (Proportional division) 35

2 **Addition and Subtraction** 38
- Exercise 2.1 (Whole numbers) 38
- Exercise 2.2 (Common fractions) 41
- Exercise 2.3 (Decimal fractions) 42

3 **Multiplication and Division** 44
- Exercise 3.1 (Rational numbers) 44
- Exercise 3.2 (Decimal fractions) 46
- Exercise 3.3 (Common fractions) 48
- Exercise 3.4 (Indices) 50
- Exercise 3.5 (Rates) 52
- Exercise 3.6 (Ratios) 53
- Exercise 3.7 (Direct proportions) 55
- Exercise 3.8 (Direct proportions) 56
- Exercise 3.9 (Direct proportions) 61
- Exercise 3.10 (Direct proportion/variation) 61

Notes 65-66

Answers i-iv

THE YOUNG SCHOLAR MATHS WORKBOOK
(NUMBERS: LEVEL 5)

ISBN: 978-0-9802922-3-7

First published: 2007

Printed in Australia by cpx printing & logistics

NUMBERS

LEVEL 5

THE YOUNG SCHOLAR

DEDICATION

This book is dedicated to the students of the authors, past and present, in Australia and in countries around the world, who benefit from their knowledge of mathematics in daily life.

ACKNOWLEDGEMENTS

The authors sincerely thank all their family members for continuing to provide encouragement, especially **Raj Aseervatham** who, despite his busy work commitments found time to render editorial advice and support to complete this book.

Illustrations: **James Hills**

Graphic Design: **Elliot Creative Design Co.** (www.elliotcreative.com)

NUMBERS

LEVEL 5

Level Statement

Students compare and order positive and negative integers, and explain and record index notation. They interpret and use conventions for expressing rates and ratios. Students use a range of computation methods to solve problems that involve positive rational numbers, rates, ratios and direct proportions.

1 Number Concepts

Core learning Outcomes

- Compare and order integers
- Add,subtract,multiply and divide integers
- Rational and irrational numbers
- Positive, negative and fractional indices
- Use and interpret rates
- Use and interpret ratios
- Use and interpret direct proportions

Integers

Integers are a set of natural numbers, their negatives and zero. All integers greater than zero are called positive integers and all integers less than zero are called negative integers. Zero and any positive integer is a whole number.

The sign '+' (plus) identifies positive integers and the sign ' - ' identifies negative integers. Zero is neither positive nor negative.

Positive and negative integers are best represented by a number line. They lie on the opposite sides of zero, positives on the right of zero and negatives on the left of zero. (The numbers are sometimes called 'directed numbers').

A positive integer such as +5 is the opposite of the negative integer -5. They are referred to as additive inverses of each other because when they are added together the answer is zero.

It is customary to write positive numbers without the sign. For example, 4 means +4 (ie: a positive number).

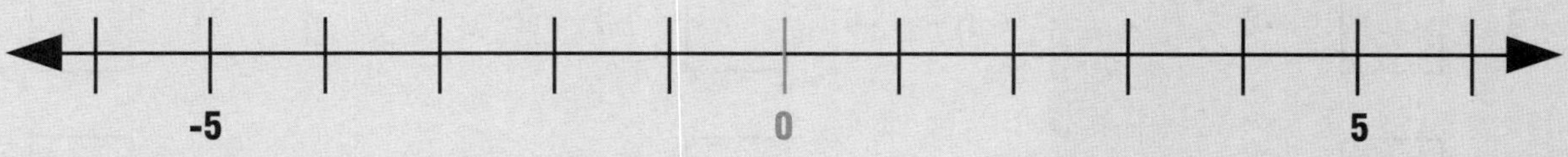

Remember: Integers consist of the positive numbers (1,2,3...), their negatives (-1, -2, -3...) and the number zero.

Absolute Value of an Integer

If we move from 0 to +5 on the number line we travel a distance of 5 places on the number line. Also if we move from +5 back to zero we travel a distance of 5 places on the number line. Although we are then back where we started, the total distance traveled on the number line is 10 and not zero!

While the total of +5 (distance traveled forwards on the number line) and -5 (distance traveled backwards on the number line) is zero, the total of the absolute values of +5 and -5 is 10.

We denote an absolute value of an integer by writing it within two vertical parallel lines as follows:

$|+5|$

The absolute value of +5 is 5.

$|-5|$

The absolute value of -5 is also 5.

We compare integers by looking at their positions on the number line. Any integer to the right of a given integer on the number line is greater than (>) that integer.

Examples: $5 > 3$ $4 > -2$ $-6 > -9$

Any integer to the left of a given integer is less than (<) that integer. Using the same numbers above we can write: $3 < 5$ $-2 < 4$ $-9 < -6$

Exercise 1.1 (Integers)

1. Write the symbol > (greater than), < (less than) or = (equal to) in the boxes. (Use the number line below to guide you.)

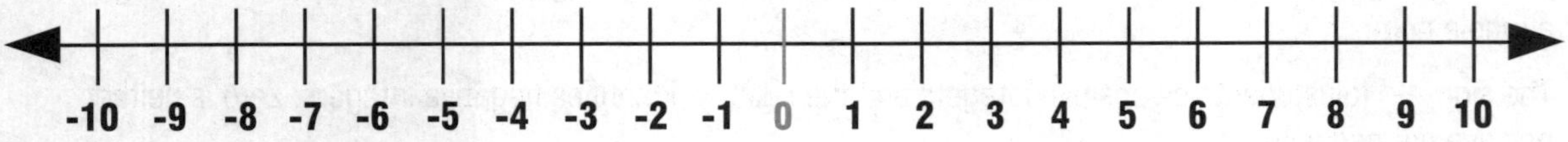

Example: -3 [>] -6

a)	**3**	□	**-2**	**e)**	**5**	□	**-8**	**i)**	**-2**	□	**-6**
b)	**9**	□	**-6**	**f)**	**-6**	□	**-2**	**j)**	**-1**	□	**5**
c)	**-10**	□	**-5**	**g)**	**3**	□	**-8**	**k)**	**-7**	□	**-4**
d)	**-8**	□	**-3**	**h)**	**7**	□	**-9**	**l)**	**0**	□	**-3**

2. Write down the absolute values of the following directed numbers.

a)	**-6**	□	**b)**	**+9**	□	**c)**	**-7**	□
d)	**15**	□	**e)**	**-11**	□	**f)**	**-13**	□

3. Write down all the integers lying between the following ranges:

a)	**-2 and 6**	□	**d)**	**-6 and 2**	□
b)	**-5 and 9**	□	**e)**	**-10 and -4**	□
c)	**0 and 10**	□	**f)**	**-3 and 3**	□

4. **Write the additive inverses of the following:**

a) -7 ☐ b) 9 ☐ c) 10 ☐

d) 8 ☐ e) -4 ☐ f) -12 ☐

5. **When moving along the number line, write down the integer you will end up with in each of the following cases (use the number line below to assist you):**

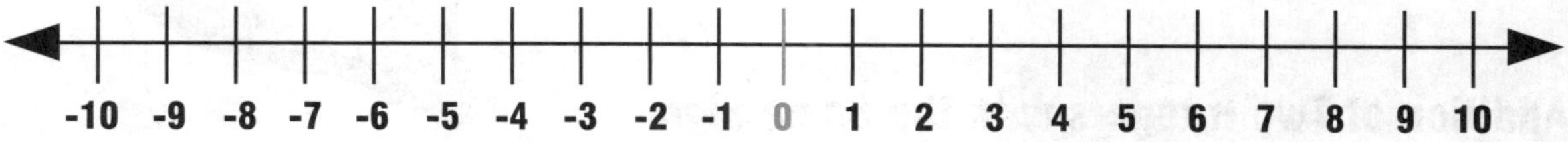

a) **5 spaces to the right of 0 first and then 3 backwards.** ☐

b) **7 spaces to the right of 0 first and then 9 backwards.** ☐

c) **8 spaces to the left of 0 and then 12 backwards.** ☐

d) **5 spaces to the left of 0 and then 10 backwards.** ☐

e) **3 spaces to the right of 0 , 5 spaces to the left and finally 8 spaces to the right.** ☐

f) **6 spaces to the left of 0, 8 spaces to the right and finally 9 spaces to the left.** ☐

6. **Write down the total distance traveled in each of the following cases while traveling along the number line.**

a) **9 spaces to the right of 0 first, and then 5 backwards.** ☐

b) **8 spaces to the right of 0 first, and then 6 backwards.** ☐

c) **1 space to the left of 0, and a further 9 left.** ☐

d) **6 spaces to the left of 0 and then 7 backwards.** ☐

e) **2 spaces to the right of 0, 7 spaces to the left and finally 5 spaces to the right.** ☐

f) **4 spaces to the left of 0, 9 spaces to the right and finally another 6 spaces to the right.** ☐

Addition of Two Integers with the Same Sign

When adding two or more integers with the same sign, simply add the absolute values and put their sign in front.

Example (i): Add +5 and + 9

Absolute values are 5 and 9 and their total = 14
Therefore, (+5) + (+9) = + 14

Example (ii): Add -6 and -7

Absolute values are 6 and 7 and their total = 13
Therefore, (-6) + (-7) = - 13

Addition of Two Integers with Opposite Signs

When adding two integers with opposite signs, simply find the difference between the absolute values and put the sign of the bigger integer in front.

Example (i): Add +3 and -5

Absolute values are 3 and 5 and their difference = 2
Therefore, (+3) + (-5) = - 2

Example (ii): Add -6 and +10

Absolute values are 6 and 10 and their difference = 4
Therefore, (-6) + (+10) = +4

When adding several integers where some are positive and others are negative, first add the integers with the same sign together and then add the resulting totals of the integers with opposite signs to get the final total.

Example: Add +7, -6, +9, -4, +2, + 8

Sum of +7, +9, +2, and +8 = +26 (Numbers with positive sign)
Sum of -6 and -4 = -10 (Numbers with negative sign)

Now we add (+26) and (-10) (Numbers with opposite signs)
Required sum= +16

Integer Subtraction

Subtracting an integer from another integer is the same as adding its additive inverse to that other integer.

Example (i): (+7) - (+5)

Additive inverse of (+5) is (-5)
So, (+7) + (-5) = +2

Example (ii): (+9) - (-3)

Additive inverse of (-3) is (+3)
So, (+9) + (+3) = 12

Example (iii): (-8) - (+4)

Additive inverse of (+4) = (-4)
So, (-8) + (-4) = -12

Example (iv): (-10) - (-2)

Additive inverse of (-2) = (+2)
So, (-10) + (+2) = -8

Exercise 1.2 (Integer addition and subtraction)

1. Write the answers for the following addition of integers in the boxes provided (use a number line):

a) (+2) + (+9) = ☐

b) (-8) + (+7) = ☐

c) (+5) + (-8) = ☐

d) (-4) + (-5) = ☐

e) (+8) + (-8) = ☐

f) (+10) + (-12) = ☐

2. Write the answers for the following subtraction of integers in the boxes provided (use a number line):

a) (+4) - (+5) = ☐

b) (-6) - (+8) = ☐

c) (-9) - (-5) = ☐

d) (-1) - (-8) = ☐

e) (+9) - (-5) = ☐

f) (-8) - (-10) = ☐

3. **Write the answers for the following addition of integers:**

a) (+12) + (+9) = ☐

b) (-16) + (+28) = ☐

c) (-14) + (-17) = ☐

d) (+21) + (-7) = ☐

e) (+15) + (+18) = ☐

f) (-31) + (-10) = ☐

4. **Write the answers for the following addition of integers:**

a) (-25) + (+17) = ☐

b) (-38) + (-22) = ☐

c) 50 + (-17) = ☐

d) 29 + (-13) = ☐

e) (+46) + (-19) = ☐

f) (-35) + (+26) = ☐

5. **Using a number line find the sum of the following integers:**

a) (+4) + (-6) + (+7) = ☐

b) (-8) + (+9) + (-3) = ☐

c) (-2) + (-4) + (-3) = ☐

d) (+8) + (-5) + (-6) = ☐

e) (-6) + (-1) + (+8) + (-1) = ☐

f) (+9) + (-5) + (7) + (+10) = ☐

6. **Write the answers for the following subtraction of directed numbers:**

a) (+14) - (-10) = ☐

b) (-19) - (-11) = ☐

c) (+25) - (-30) = ☐

d) (+65) - (-35) = ☐

e) (-100) - (+150) = ☐

f) (-120) - (-38) = ☐

Multiplication of Integers

When multiplying integers, the sign the product takes depends on the sign of the individual integer. The rule is that integers with like signs give a product with a positive sign. Integers with opposite signs always give a product with a negative sign.

(+)	x	(+)	=	(+)
(-)	x	(-)	=	(+)
(+)	x	(-)	=	(-)
(-)	x	(+)	=	(-)

Examples:	Multiplying	+8	by	+6	gives	+48
	Multiplying	-8	by	-6	gives	+48
	Multiplying	+8	by	-6	gives	-48
	Multiplying	-8	by	+6	gives	-48

The same rule is extended when more than two integers are multiplied together.

Example: Multiply +2, -3 and +4

Multiplying	+2	by	-3	gives	-6
Multiplying	-6	by	+4	gives	-24

Dividing an Integer by Another Integer

A similar rule is applied when one integer is divided by another.

The rule is that integers with like signs give a quotient with a positive sign. Integers with opposite signs always give a quotient with a negative sign.

(+)	÷	(+)	=	(+)
(-)	÷	(-)	=	(+)
(+)	÷	(-)	=	(-)
(-)	÷	(+)	=	(-)

Examples:	Dividing	+18	by	+6	gives	+3
	Dividing	-18	by	-6	gives	+3
	Dividing	+18	by	-6	gives	-3
	Dividing	-18	by	+6	gives	-3

Exercise 1.3 (Integer multiplication)

1. **Write the answers for the following multiplication of integers in the boxes provided:**

a) (+3) x (+4) = ☐

b) (+5) x (-6) = ☐

c) (-6) x (-7) = ☐

d) (-9) x (+8) = ☐

e) (10) x (-4) = ☐

f) (11) x (-5) = ☐

2. **Write the answers for the following division of integers in the boxes provided:**

a) (+8) ÷ (+2) = ☐

b) (-6) ÷ (-3) = ☐

c) (+12) ÷ (-4) = ☐

d) (-10) ÷ (+5) = ☐

e) (14) ÷ (-7) = ☐

f) (-16) ÷ (-4) = ☐

3. **Write the answers for the following multiplication of integers in the boxes provided:**

Example:

(+3) x (-2) x (-7) = (-6) x (-7) = +42

a) (+2) x (-4) x (-3) = ☐ = ☐

b) (+3) x (+5) x (-2) = ☐ = ☐

c) (-4) x (-2) x (-3) = ☐ = ☐

d) (+6) x (-4) x (+2) = ☐ = ☐

e) (-5) x (-4) x (+7) = ☐ = ☐

f) (8) x (3) x (-4) = ☐ = ☐

4. **Write the answers for the following division of integers in the boxes provided (do the division in the given order):**

Example:

(-12) ÷ (-2) ÷ (+3)	=	(+6) ÷ (+3)	=	+2

a)	**(+9) ÷ (+3) ÷ (-3)**	=		=	
b)	**(+8) ÷ (+4) ÷ (-1)**	=		=	
c)	**(-12) ÷ (-2) ÷ (-3)**	=		=	
d)	**(+10) ÷ (-1) ÷ (+5)**	=		=	
e)	**(-15) ÷ (-5) ÷ (+1)**	=		=	
f)	**(20) ÷ (-2) ÷ (-5)**	=		=	

5. **Write the answers for the following mixed multiplication and division of integers in the boxes provided (do the calculations in the given order):**

a)	**(+6) x (+3) ÷ (-2)**	=		=	
b)	**(+8) ÷ (-2) x (-9)**	=		=	
c)	**(-7) x (-8) ÷ (-4)**	=		=	
d)	**(6) x (9) ÷ (-3)**	=		=	
e)	**(-12) ÷ (-6) x (-7)**	=		=	
f)	**(-10) x (5) ÷ (-2)**	=		=	

6. **Write the answers for the following mixed integer multiplication and division in the boxes provided (do the division first):**

a) **(-9) x (+8) ÷ (-4)** = ☐ = ☐

b) **(-12) ÷ (-6) x (-3)** = ☐ = ☐

c) **(-20) ÷ (-10) x (-2)** = ☐ = ☐

d) **(+27) ÷ (-9) x (-2)** = ☐ = ☐

e) **(-36) ÷ (-12) x (-8)** = ☐ = ☐

f) **(7) x (-42) ÷ (-6)** = ☐ = ☐

Square Root of Numbers

Consider the square number 4. It can be written as 2 x 2 or -2 x -2

The numbers 2 and -2 are called the square roots of 4.

Take another square number 49. It can be written as 7 x 7 or -7 x -7.

The numbers 7 and -7 are called the square roots of 49.

Only the square numbers will give exact square roots.

The square root of 4 is written as $\sqrt{4} = \pm 2$ (positive or negative 2)

The square root of 49 is written as $\sqrt{49} = \pm 7$ (positive or negative 7)

The sign "$\sqrt{\ }$" is referred to as 'surd'. When written as $\sqrt[2]{\ }$ it means square root. Generally the 2 is left out and the sign "$\sqrt{\ }$" means square root.

Calculating Square Roots of Numbers

Example (i): Calculate the square root of 4.

Find a number that when multiplied by itself gives a number very close to 4 or equal to 4. In this case it is 2. Write the '2' on either side of the vertical parallel lines. Now multiply the two 2's together and write the answer 4 under the original number 4 and subtract. If the answer is zero, then the number on the right of the parallel lines is the exact positive square root of 4.

2	4	2	⟶ Square root
	4		
	0		

One square root of 4 is $+2$ but it can also be -2 as both $+2$ and -2 when squared gives 4. Hence, $\sqrt{4} = \pm 2$ (± 2 is pronounced 'plus or minus 2')

Example (ii): Calculate the square root of 49.

7	49	7	⟶ Square root
	49		
	0		

$\sqrt{49} = \pm 7$

When the number to be square rooted has more than two digits, divide it in pairs starting from the right end of the number. Sometime we will be left with a single digit number on the left and other times a two digit number.

Example (iii): Calculate the square root of 289.

Find a number that when multiplied by itself gives a number very close to or equal to the single digit 2. In this case it is 1. Write the '1' on either side of the vertical parallel lines. Now multiply the two '1's together and write the answer 1 under the number 2 and subtract. The result is 1.

1	2.89	1
	1	
	1	

Bring down the pair 89. Double the number '1' on the right of the parallel line and write the result 2 on the left of the parallel line.

1	2.89	1
	1	
2	1 89	

Now choose and write a suitable number (in this case 7) both next to "1' on the right of the parallel lines and next to '2' on the left of the parallel such that when the new number on the left of the parallel line (in this case 27) is multiplied by the chosen suitable number 7 gives a result closer to (and under 189) or equal to 189. Write it down and subtract from the original 189. If the answer is zero, we have got the exact square root of 289.

1	2.89	17	⟶ Square root
	1		
27	189	17	
	189		
	0		

$\sqrt{289} = \pm 17$

Calculating Square Roots of Non-Square Numbers

Example (i): Find the square root of 24.

24 being a non-square number, its square root is not going to be a whole number. We first add a number of zeros after the decimal point at the end of 24 and divide them in pairs.

The number to write first on either side of the parallel line is 4.

4 multiplied by 4 gives the closest number to 24 (and under 24).

We follow the same procedure as with square numbers. Every time we bring down a pair of zeros, we double the number that remains on the right of the parallel lines and write the result on the left of the parallel line as shown. The number that is to be doubled is 4 and the answer 8 is written on the left side of the parallel line. Place the decimal point after the '4' on the right of the parallel line.

Now, find a suitable number and write it next to both 4 (on the right of parallel line) and 8 (on the left of the parallel line). This number is 8. Multiply the '88' by the 8 written last on the right of the parallel line to get 704) Any number chosen other than 8 will not give the desired product which has to be as close to 800 as possible and not exceed 800. For example , if we chose 9, then we will be multiplying 89 by 9 and get 801! If we chose 7, it won't give a product as close to 800 as 8 will give.

Continue this process to get the desired number of decimal places for the answer.

```
  4 | 24.00,00,00 | 4.89
    | 16          |
 88 |   800       |
    |   704       |
969 |    9600     |
    |    8721     |
    |     879     |
```

$\sqrt{24} = \pm 4.89 = 4.9$ (corrected to 1 decimal place)

Rational and Irrational Numbers

Numbers can be categorised as follows:

- real numbers (any number on the number line)
- complex numbers (numbers with real and imaginary parts)

Real numbers are further categorised as:

- rational numbers
- irrational numbers

Rational numbers are all the numbers we have come across so far. They include whole numbers, integers and both positive and negative common fractions, decimal fractions and mixed numbers.

Examples: $4, -9, 6.5, -\frac{1}{3}, 0, 3.4568$

Generally any number that can be expressed as a common fraction is a rational number. For example 4 is rational because it can be expressed in fractional form as $\frac{4}{1}$. Similarly a number such as -9 is a rational number. 0 is also a rational number ($\frac{0}{1}$).

Square roots of square numbers are also rational.

Examples: $\sqrt{4}$, $\sqrt{49}$

A terminating decimal fraction is a rational number.

Example: 2.75 (expressed as a common fraction is $2\frac{3}{4}$)

A non-terminating decimal fraction with repeating digit(s) that can be expressed as a proper fraction is rational.

Examples:

$0.333333333......-- = \frac{3}{9} = \frac{1}{3}$

$0.090909...... = \frac{9}{99} = \frac{1}{11}$

$0.212121...... = \frac{21}{99} = \frac{7}{33}$

Any number that is not rational is irrational. Generally square roots of non-square numbers are irrational.

Examples: $\sqrt{2}$, $\sqrt{3}$ $\sqrt{8}$ (These are all non-terminating decimal fractions and hence irrational)

We can now categorise numbers as real, rational, irrational and integers.

Examples:

Examples:	7	Real, rational, integer
	$\sqrt{8}$	Real, irrational
	-5	Real, rational, integer

Square roots of negative numbers are complex numbers.

Examples: $\sqrt{-1}$ $\sqrt{-6}$

Exercise 1.4 (Square numbers and square roots)

1. **Underline the square number in each of the following sets of numbers.**

a) 2 7 9 12

b) 6 15 19 25

c) 10 14 16 32

d) 17 36 42 51

e) 35 55 64 79

f) 81 95 102 135

2. **Write down the square roots of the following common square numbers:**

a) 9 ☐

b) 16 ☐

c) 25 ☐

d) 36 ☐

e) 64 ☐

f) 81 ☐

3. **Write down the square roots of the following three digit square numbers by inspection:**

a) 100 ☐

b) 121 ☐

c) 169 ☐

d) 196 ☐

e) 225 ☐

f) 324 ☐

4. **Complete the process and find the positive square roots of the following non-square numbers correct to 2 decimal places.**

a)

1	2.00,00,00	1.4
	1	
24	100	
	96	
28	400	

$\sqrt{2}$ = ____

b)

1	3.00,00,00	1.7
	1	
27	200	
	189	
	1100	

$\sqrt{3}$ = ____

c)

2	5.00,00,00	2.2
	4	
42	100	
	84	
44	1600	

$\sqrt{5}$ = ____

5. **Write down the approximate positive square roots of the following numbers to the nearest whole number.**

a) 15 ____

b) 26 ____

c) 50 ____

d) 66 ____

e) 83 ____

f) 99 ____

6. **Use a calculator to find the positive square roots of the following correct to three decimal places.**

a) 15 ☐ **d)** 66 ☐

b) 26 ☐ **e)** 83 ☐

c) 50 ☐ **f)** 99 ☐

Exercise 1.5 (Rational and irrational numbers)

1. **Write the following rational numbers as <u>common fractions</u> in simplest form.**

a) 0.2 ☐ **d)** $0.\overline{33}$ ☐

b) -0.08 ☐ **e)** 2.375 ☐

c) -3.5 ☐ **f)** $\sqrt{1/9}$ ☐

2. State whether the following numbers are rational or irrational.

a) $\frac{2}{5}$

d) $\sqrt{16}$

b) $\sqrt{12}$

e) $-\frac{1}{8}$

c) -9

f) $0.\overline{27}$

3. Classify the following numbers into real and rational, irrational or integer, or complex.

a) 3.14

d) $\sqrt{10}$

b) $\sqrt{25}$

e) $0.\overline{66}$

c) 3/14

f) $\sqrt{-36}$

4. Place a tick [✓]in the box if the number is irrational. Otherwise place a cross [✗] in the box.

a) $\sqrt{13}$

b) $\sqrt{49}$

c) $\sqrt{60}$

d) $\sqrt{81}$

e) $\sqrt{75}$

f) $\sqrt{33}$

5. State whether each of the following rational numbers is a terminating decimal or a recurring decimal.

a) $\frac{3}{4}$

d) $\frac{2}{3}$

b) $\frac{2}{7}$

e) $\frac{7}{8}$

c) $\frac{3}{8}$

f) $\frac{1}{11}$

6. Place a tick [✓] in the box if the number is rational. Otherwise place a cross [✗] in the box.

a)	**0.2275**	☐	**b)**	**0.22222...**	☐	**c)**	**0.393939...**	☐
d)	**0.666666...**	☐	**e)**	**$0.\overline{102}$**	☐	**f)**	**2.2360679...**	☐

Raising a NUMBER to a Power

When a number is multiplied by itself, the resulting number is a square number. For example 2 x 2 = 4 and so 4 is a square number.

2 x 2 can be written as 2^2 and is read "2 squared". or 2 raised to the power 2.

3 x 3 x 3 can be written as 3^3 and is read "3 cubed" or, 3 raised to the power 3.

The 'power' is the number to which a given base number is raised. The power is commonly referred to as the *index* or *exponent*.

The index can be a positive or negative whole number or a positive or negative fraction.

Examples of numbers with a positive or negative whole number index: 5^3, 3^4, 8^{-2}, 6^{-3}, 10^{-5}

Examples of numbers with a positive or negative fractional index: $3^{1/2}$, $4^{1/3}$, $9^{-1/2}$, $32^{1/5}$, $16^{-1/4}$

Each of the above has a meaning as follows:

$3^{1/2}$ = Square root of 3 = $\sqrt{3}$

$4^{1/3}$ = Cube root of 4 = $\sqrt[3]{4}$

$9^{-1/2}$ = Square root of inverse of 9 = $1/\sqrt{9}$

$32^{1/5}$ = Fifth root of 32 = $\sqrt[5]{32}$

$16^{-1/4}$ = Fourth root of inverse of 16 = $1/\sqrt[4]{16}$

A negative index can be made positive and a positive index made negative by inverting the base number.

For example 3^{-2} is written as $\frac{1}{3^2}$ and $\frac{1}{3^{-2}}$ is written as 3^2.

Exercise 1.6 (Indicies)

1. Write the following in index form.

Examples: **2 x 2 x 2 x 2 = 2^4** **4 x 4 x 4 x 5 x 5 = $4^3 \times 5^2$**

a)	**2 x 2 x 2 =**	☐	**d)**	**6 x 6 x 6 =**	☐
b)	**3 x 3 x 3 x 3 =**	☐	**e)**	**2 x 2 x 2 x 3 x 3 =**	☐
c)	**5 x 5 x 5 x 5 x 5 =**	☐	**f)**	**10 x 10 x 10 x 10 =**	☐

2. **Write each of the following indices in expanded form and find its value.**

a) 3^5 = =

b) 2^6 = =

c) 5^4 = =

d) 10^6 = =

e) $2^4 \times 10^2$ = =

f) $4^3 \times 2^2$ = =

3. **Convert the positive index to a negative index and vice versa.**

a) 2^3

b) 3^1

c) 7^{-4}

d) 10^4

e) 5^{-6}

f) 3^{-5}

4. **Write the index form of the following surds.**

a) $\sqrt{4}$

b) $\sqrt{9}$

c) $\sqrt[3]{27}$

d) $\sqrt[5]{32}$

e) $\sqrt{16}$

f) $\sqrt[4]{81}$

5. Simplify the following:

a) $\sqrt{16}$

b) $\sqrt[3]{64}$

c) $\sqrt[4]{625}$

d) $\sqrt[5]{32}$

e) $\sqrt[4]{100^2}$

f) $(\sqrt[3]{8})^2$

6. Use a calculator and find the answer for the following:

a) $\sqrt{81}$

b) $\sqrt[3]{64}$

c) $\sqrt[4]{256}$

d) $\sqrt[5]{243}$

e) $(3^2)^{-2}$

f) $(6^4)^{\frac{1}{2}}$

Rate

The word 'rate' is used to signify a quantity in relation to a unit of another quantity. This unit of quantity can be measured as a unit of time (second, minute or hour), volume (litre or millilitre), manufacture, sale, currency and so on.

The common rate calculations we encounter in real life are:

- labour rate
- charge out rate
- rate of return on investment
- rate of interest
- rate of flow
- rate of travel (speed)
- rate of change
- price per unit
- cost per unit
- heart beat rate
- money exchange rate

Example (i): John was paid $300 for work he performed for 5 hours. What is the average hourly rate of pay?

Hourly rate of pay = $300/5 = $60 per hour

Example (ii): Convert a speed (travel rate) of 18 km per hour into metres per second.

In 1 hour distance travelled = 18 km

In 3 600 seconds distance travelled = 18 000 m

Distance travelled in 1 second = 18 000 metres / 3 600 seconds = 5 m / sec.

Example (iii): If 64 Australian dollars can be exchanged for 50 US dollars, what is the exchange rate of A$1 to US$1?

A$64 = US$ 50

A$1 = 50/64 = US$0.78

Exercise 1.7 (Rates)

1. If Anne earns $450 for 10 hours of work, what is her hourly rate of pay?

Hourly rate of pay = $\frac{\text{earnings}}{\text{hours}}$ = ______ = $ [ ] per hour

2. If Peter can run 100 metres in 10 seconds, calculate his speed in metres per second.

Speed = $\frac{\text{meters}}{\text{seconds}}$ = ______ = [ ] meters per second

3. If it takes 20 minutes to travel 30 km by car, What is the speed of the car in km per hour?

Speed = $\frac{\text{km}}{\text{hours}}$ = ______ = [ ] km per hour

4. A car loses its value at the rate of $5 600 in 5 years, What is the rate of value loss (depreciation) per year?

Depreciation = $\frac{\text{loss}}{\text{years}}$ = ______ = $ [ ] per year

5. A factory produces 700 toy cars in $1\frac{3}{4}$ hours, What is the rate of production per hour?

Rate of production = $\frac{\text{toy cars}}{\text{hours}}$ = ______ = [ ] toy cars per hour

6. If A$625 can be exchanged for £250 in the United Kingdom, what is the exchange rate of an Australian dollar to the United kingdom currency?

A$ 625 = £250

A$ 1 = ______

= £ [ ]

Exercise 1.8 (Rates)

1. A car park charges $12 for the first hour of parking and $8 for each additional half-hour. If a person parks the car for 2 hours, what is the average parking rate per hour he pays?

Average parking rate = $\frac{\text{dollars}}{\text{hours}}$ = ______ = $ ______ dollars per hour

2. After covering half the distance, a driver finds that there is another 200 km left to reach the destination. If the total time of travel is 4 hours, what is the average rate (speed) of travel?

Speed = ______ = ______ = ______ km per hour

3. A water tank with a capacity of 300 litres is filled in half an hour by a tap from which water is flowing at a steady rate. What is the rate of flow of water in litres per minute?

Rate of water flow = $\frac{\text{litres}}{\text{minutes}}$ = ______ = ______ litres per minute

4. A person borrowed $3 000 for one year from a bank and paid simple interest of $180. What is the yearly rate of interest he paid per dollar? What is the yearly interest rate as a percentage?

Yearly interest rate = $\frac{\text{interest}}{\text{borrowing}}$ = ______ = ______ cents per dollar

= ______ %

5. By investing $10 000 in a long term project, William Smith earned a return of $1 000 in the first year. What rate of return per dollar of investment did William Smith receive in the first year? What is the rate of return as a percentage of the investment?

First year rate of return = $\frac{\text{return}}{\text{investment}}$ = ______ = ______ cents per dollar

= ______ %

6. If A$950 can be exchanged for C$1 150 in Canada, what is the exchange rate of an Australian dollar to the Canadian dollar?

A$ 950 = C$1 150

A$ 1 = C$ ______

= C$ ______

Ratios

A way of stating the relationship between two quantities is called a ratio. For example, if there are twice as many girls in a class as boys, we state this relationship as:

girls to boys = 2 to 1

It is customary to write this as: girls to boys = 2 : 1 or $\frac{2}{1}$

Ratios are usually stated as whole number comparisons. e.g. 2 to 1, 2 to 3, 4 to 9 etc.

Equivalent Ratios

A ratio such as 2 : 1 when multiplied by any number will remain the same:

That is, 2 : 1 = 4 : 2 = 6 : 3 = 8 : 4 and so on.

They are all equivalent ratios. 4 to 2, 6 to 3 , 8 to 4 etc. are all equivalent to2 to 1.

2 to 1 is the simplest of the equivalent forms.

For the above example, the ratio of girls to total number of students in the class is 2 to 3.

Similarly, the ratio of boys to the total number of students in the class is 1 to 3.

Spotting an equivalent ratio

The method of finding out whether two ratios are equivalent or not is to carry out a 'cross multiplication' as shown below. If the two products are the same, the ratios are equivalent, otherwise they are not.

Example (i): Are the two ratios 2 to 3 and 4 to 6 equivalent?

$\frac{2}{3} \times \frac{4}{6}$

Cross multiplication: 2 x 6 is equal to 3 x 4

The two ratios are thus equivalent.

Example (ii): Are the two ratios 2 to 5 and 3 to 10 equivalent?

$\frac{2}{5} \times \frac{3}{10}$

Cross multiplication: 2 x 10 is not equal to 5 x 3

The two ratios are not equivalent.

Ratios of Fractions

Example (i): Find the ratio of $\frac{2}{9}$ to $\frac{1}{6}$.

First find the LCD of 9 and 6. This is 18.

Multiply the two given fractions by the LCD.

This gives the ratio: $\frac{2}{9} \times 18$ to $\frac{1}{6} \times 18 = 4 : 3$

Example (ii): Find the ratio of $1\frac{1}{2}$ to $2\frac{1}{4}$.

This is the same as $\frac{3}{2}$ to $\frac{9}{4}$.

First find the LCD of 2 and 4. This is 4.

Multiply the two given fractions by the LCD.

This gives the ratio: $\frac{3}{2} \times 4$ to $\frac{9}{4} \times 4 = 6 : 9 = 2 : 3$ (simplest form)

Example (iii): Find the ratio of 1.25 to 0.75.

Multiplying by 100, ratio = 125 : 75

Dividing by 25, ratio = 5 : 3 (ie: divide by 5 first and by 5 again)

Ratios of Lengths, Areas, Volumes and Weights

When comparing two or more lengths, two or more areas, volumes or weights it is necessary that they are all stated in the same units (usually the smaller units).

Example (i): Find the ratio of 25 cm to 1 metre.

Ratio = 25 cm : 100 cm
(Centimetre is the smaller unit) = 1 : 4

Example (ii): Find the ratio of 500 cm^2 to 1 m^2.

Ratio = 500 cm^2 : 10 000 cm^2 = 1 : 20

Example (iii): Find the ratio of 200 ml to 1 litre

Ratio = 200 ml : 1 000 ml = 1 : 5

Example (iv): Find the ratio of 600 grams to 2 kg

Ratio = 600 grams : 2 000 grams = 3 : 10

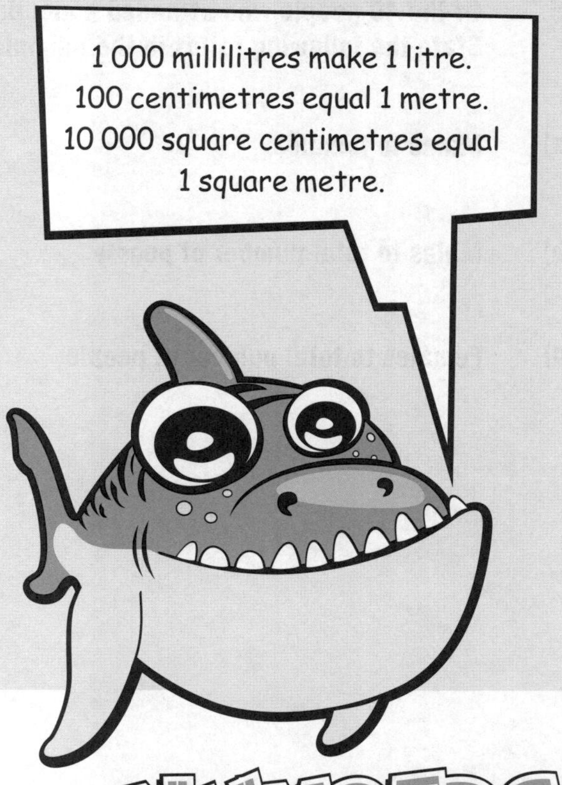

Exercise 1.9 (Ratios)

1. You have three pens and two pencils. What is the ratio of pens to pencils?

Pens : Pencils =

2. A boy bought nine trousers and seven shirts. What is the ratio of trousers to shirts?

Trousers : Shirts =

3. A girl has eight fifty dollar notes and three twenty dollar notes.
What is the ratio of fifty dollar notes to twenty dollar notes?

Fifty dollar notes: Twenty dollar notes =

4. Write the following ratios in their simplest form:

a) 10 : 8 b) 25 : 15 c) 12 : 16

d) 28 : 35 e) 72 : 45 f) 84 : 36

5. Of the 40 people who attended a meeting there were 24 males and 16 females.
State the following ratios in their simplest form.

a) Males to females =

b) Males to total number of people =

c) Females to total number of people =

6. In a box there are 12 green balls and 18 red balls. State the following ratios in their simplest form.

a) Green balls to red balls = [ ]

b) Red balls to green balls = [ ]

c) Green balls to total number of balls = [ ]

d) Red balls to total number of balls = [ ]

Exercise 1.10 (Ratios)

1. Write two other simpler equivalent ratios for each of the following:

a) 4 : 8 = [ ] = [ ]

b) 12 : 18 = [ ] = [ ]

c) 8 : 20 = [ ] = [ ]

d) 18 : 42 = [ ] = [ ]

e) 36 : 30 = [ ] = [ ]

f) 66 : 42 = [ ] = [ ]

2. Write the following ratios in their simplest form:

a) 2 : 4 : 8 = [ ]

b) 12 : 18 : 24 = [ ]

c) 16 : 24 : 40 = [ ]

d) 20 : 25 : 30 = [ ]

e) 2.5 : 3 : 3.5 = [ ]

(Multiply each number by 10)

= [ ]

f) 4.8 : 3.2 : 8 = [ ]

(Multiply each number by 10)

= [ ]

3. Write the following ratios in their simplest form:

a) $1\frac{1}{2} : 3$ = [ ] = [ ]

b) $3\frac{1}{4} : 2\frac{3}{5}$ = [ ] = [ ]

c) $4\frac{2}{3} : 3\frac{1}{2}$ = [ ] = [ ]

d) $6\frac{3}{5} : 3\frac{3}{10}$ = [ ] = [ ]

e) $2\frac{1}{4} : 3\frac{3}{8} : 4\frac{1}{2}$ = [ ] = [ ]

f) $5\frac{1}{2} : 4\frac{2}{5} : 6\frac{3}{5}$ = [ ] = [ ]

4. Place a tick [✓] in the box, if in each case the given ratios are the same. Otherwise place a cross [✗] in the box.

a) 1 : 2 and 5 : 10 [ ]

b) 3 : 4 and 9 : 11 [ ]

c) 4 : 9 and 12 : 24 [ ]

d) 7 : 8 and 21 : 24 [ ]

e) 9 : 12 and 12 : 16 [ ]

f) 1 : 6 and 8 : 40 [ ]

5. Write the following ratios in their simplest form. Pay attention to the different units.

a) 40 cm : 1 m : 60 cm [ ]

b) 1.5 kg : 500g : 1 kg [ ]

c) 1.2 litres : 600ml : 0.9 litres [ ]

d) 2 hours : 90 min. : 75 min. [ ]

e) $5 : 200 c : $1.50 [ ]

f) 2 years : 26 weeks : 91 days [ ]

6. **Write in the box the larger of the two given ratios (use a calculator if necessary).**

a) 3 :4 and 4 : 5

b) 2 : 9 and 1 : 6

c) 1 : 8 and 1 : 11

d) 7 : 12 and 5 : 9

e) 5 : 14 and 3 : 8

f) 19 : 25 and 17 : 22

Direct Proportion

A relationship between two quantities such that when one increases the other. increases and when one decreases the other decreases is called direct proportion.

Two quantities in direct proportion increase or decrease in the same ratio as each other.

Example (i):	1 ice-cream costs	\$3
	2 ice-creams cost	\$6
	3 ice-creams cost	\$9

We observe that when the number of ice-creams is doubled (from 1 to 2), the cost also doubles (from \$3 to \$6). The ratio 1:2 represents the increase in number of ice creams and is the same as the ratio 3:6 which represents the increase in the cost.

When the number of ice-creams is tripled (from 1 to 3), the cost also triples (from \$3 to \$9). The ratio 1:3 is the same as the ratio 3:9.

The two quantities, ice-creams and cost, increase in the same ratio as each other. They are in direct proportion.

Example (ii):	12 identical bricks weigh	36 kg
	6 identical bricks weigh	18 kg
	3 identical bricks weigh	9 kg

We observe that when the number of bricks is halved (from 12 to 6), the weight also halves (from 36 kg to 18 kg). The ratio 12:6 is the same as the ratio 36:18 (both simplify to 2:1).

When the number of bricks is reduced to a quarter (from 12 to 3), the weight is also reduced to a quarter (from 36 to 9). The ratio 12:3 is the same as the ratio 36:9 (both simplify to 4:1).

The two quantities, bricks and weight decrease in the same ratio as each other. They are in direct proportion.

If two ratios of two quantities are equivalent then the two quantities are in direct proportion and the cross product of the two ratios will be equal.

Example: $\frac{2}{3} \quad \frac{6}{9}$

Then $2 \times 9 = 6 \times 3$

The two equivalent ratios 2/3 and 6/9 can be written as:

2:3 :: 6:9

This is read: 2 is to 3 is as 6 is to 9.

The statement 2 shirts cost $40 and 4 shirts cost $80 can be written as:

2:4 :: 40:80

We observe that 4 is twice 2 and 80 is twice 40

Example (i): Find the unknown if 1:8 :: ?:24

Since 24 is three time 8, the unknown must be 3 x 1 = 3

Example (ii): Find the unknown if 18:45 :: 2:?

Since 2 is a ninth of 18, the unknown must be a ninth of 45 =5

Exercise 1.11 (Direct proportion)

1. Place a tick [✓] in the box if a pair of ratios is equivalent, otherwise place a cross [✗] in the box.

a) 3 : 5 and 12 : 20 ☐

b) 4 : 9 and 20 : 45 ☐

c) 5 : 10 and 10 : 30 ☐

d) 18 : 27 and 4 : 9 ☐

e) 20 : 35 and 4 : 7 ☐

f) 32 : 48 and 2 : 3 ☐

NUMBERS

LEVEL 5

Answers

LEVEL 5

Exercise 1.1

1. (a) 3>-2 (b) 9>-6 (c) -10<-5 (d) -8<3 (e) 5>-8
   (f) -6 <- 2 (g) 3>-8 (h) 7>-9 (i) -2>-6 (j) -1<5
   (k) -7<-4 (l) 0>-3
2. (a) 6 (b) 9 (c) 7 (d) 15 (e) 11 (f) 13
3. (a) -1, 0, 1, 2, 3, 4, 5
   (b) -4, -3, -2, -1, 0, 1, 2, 3, 4, 5, 6, 7, 8
   (c) 1, 2, 3, 4, 5, 6, 7, 8, 9
   (d) -5, -4, -3, -2, -1, 0, 1
   (e) -9, -8, -7, -6, -5
   (f) -2, -1, 0, 1, 2
4. (a) +7 (b) -9 (c) -10 (d) -8 (e) +4 (f) +12
5. (a) +2 (b) -2 (c) +4 (d) +5 (e) +6 (f) -7
6. (a) 14 (b) 14 (c) 10 (d) 13 (e) 14 (f) 19

Exercise 1.2

1. (a) +11 (b) -1 (c) -3 (d) -9 (e) 0 (f) -2
2. (a) -1 (b) -14 (c) -4 (d) +7 (e) +14 (f) +2
3. (a) +21 (b) +12 (c) -31 (d) +14 (e) +33 (f) -41
4. (a) -8 (b) -60 (c) +33 (d) +16 (e) +27 (f) -9
5. (a) +5 (b) -2 (c) -9 (d) -3 (e) 0 (f) + 21
6. (a) +24 (b) -8 (c) +55 (d) +100 (e) -250 (f) -82

Exercise 1.3

1. (a) +12 (b) -30 (c) +42 (d) -72 (e) -40 (f) -55
2. (a) +4 (b) +2 (c) -3 (d) -2 (e) -2 (f) +4
3. (a) +24 (b) -30 (c) -24 (d) -48 (e) +140 (f) -96
4. (a) -1 (b) -2 (c) -2 (d) -2 (e) +3 (f) +2
5. (a) -9 (b) +36 (c) -14 (d) -18 (e) -14 (f) +25
6. (a) +18 (b) -6 (c) -4 (d) +6 (e) -24 (f) +49

Exercise 1.4

1. (a) 9 (b) 25 (c) 16 (d) 36 (e) 64 (f) 81
2. (a) ±3 (b) ±4 (c) ±5 (d) ±6 (e) ±8 (f) ±9
3. (a) ±10 (b) ±11 (c) ±13 (d) ±14 (e) ±15 (f) ±18
4. (a) ±1.41 (b) ±1.73 (c) ±2.24
5. (a) 4 (b) 5 (c) 7 (d) 8 (e) 9 (f) 10
6. (a) 3.873 (b) 5.099 (c) 7.071 (d) 8.124 (e) 9.110 (f) 9.950

Exercise 1.5

1. (a) 1/5 (b) -2/25 (c) -7/2 (d) 1/3 (e) 19/8 (f) ±1/3
2. (a) rational (b) irrational (c) rational (d) rational (e) rational
   (f) rational
3. (a) real, rational (b) real, rational, integer
   (c) real rational (d) real, irrational
   (e) real, rational (f) complex
4. (a) ✓ (b) ✗ (c) ✓ (d) ✗ (e) ✓ (f) ✓
5. (a) terminating (b) recurring (c) terminating
   (d) recurring (e) terminating (f) recurring
6. (a) ✓ (b) ✓ (c) ✓ (d) ✓ (e) ✓ (f) ✗

Exercise 1.6

1. (a) 2^3 (b) 3^4 (c) 5^5 (d) 6^3 (e) $2^3 \times 3^2$ (f) 10^4
2. (a) 243 (b) 64 (c) 625 (d) 1 000 000 (e) 1 600 (f) 256
3. (a) $1/2^{-3}$ (b) $1/3^{-1}$ (c) $1/7^4$ (d) $1/10^{-4}$ (e) $1/5^6$ (f) $1/3^5$
4. (a) $4^{1/2}$ (b) $9^{1/2}$ (c) $27^{1/3}$ (d) $32^{1/5}$ (e) $16^{1/2}$ (f) $81^{1/4}$
5. (a) 4 (b) 4 (c) 5 (d) 2 (e) 10 (f) 4
6. (a) 9 (b) 4 (c) 4 (d) 3 (e) 1/81 (f) 36

Exercise 1.7

1. $45 per hour
2. 10 m/s
3. 90 km per hr.
4. $1 120 per year
5. 400 toy cars per hr.
6. A$1 = £0.40

Exercise 1.8

1. $14 per hour
2. 100 km per hr
3. 10 litres/min.
4. 0.06 cents per dollar; 6%
5. 10 cents per dollar; 10%
6. A$1 = C$ 1.21

Exercise 1.9

1. 3: 2
2. 9:7
3. 8:3
4. (a) 5:4 (b) 5:3 (c) 3:4 (d) 4:5 (e) 8:5 (f) 7:3
5. (a) 3:2 (b) 3:5 (c) 2:5
6. (a) 2:3 (b) 3:2 (c) 2:5 (d) 3:5

Exercise 1.10

1. (a) 1:2; 2:4 (b) 2:3; 4:6 (c) 2:5; 4:10
   (d) 3:7; 6:14 (e) 6:5; 12:10 (f) 11:7; 22:14
2. (a) 1:2:4 (b) 2:3:4 (c) 2:3:5 (d) 4:5:6 (e) 5:6:7 (f) 3:2:5
3. (a) 1:2 (b) 5:4 (c) 4:3 (d) 2:1 (e) 2:3:4 (f) 5:4:6
4. (a) ✓ (b) ✗ (c) ✗ (d) ✓ (e) ✓ (f) ✗
5. (a) 2:5:3 (b) 3:1:2 (c) 4:2:3 (d) 8:6:5 (e) 10:4:3 (f) 8:2:1
6. (a) 4:5 (b) 2:9 (c) 1:8 (d) 7:12 (e) 3:8 (f) 17:22

Exercise 1.11

1. (a) ✓ (b) ✓ (c) ✗ (d) ✗ (e) ✓ (f) ✓
2. (a) ✗ (b) ✓ (c) ✓ (d) ✗ (e) ✓ (f) ✓
3. (a) ✓ (b) ✗ (c) ✗ (d) ✓ (e) ✗ (f) ✓
4. (a) 21 (b) 15 (c) 2 (d) 1 (e) 10 (f) 1.2
5. 50 litres
6. $6

Exercise 1.12

1. (a) Abel, $75; Arvo, $100
2. Partner 1, $2 000; Partner 2, $1 200
3. Peta,$3 120; Stacey, $2 730;Elizabeth,$1 950
4. A, 420g; B, 280g; C, 140g
5. Tin, 2 kg; Copper, 1.5 kg; Silver 5 kg
6. Profit, $12 150

Exercise 2.1

1. 366
2. $1 111
3. $11 478, $11 402
4. 4 128 kg
5. 21 435 000
6. $1 730 000

Exercise 2.2

1. $1\frac{13}{24}$
2. $\frac{6}{7}$
3. $32\frac{41}{60}$
4. $3\frac{3}{8}$
5. $1\frac{5}{24}$ kg
6. $1\frac{11}{12}$ kg

Exercise 2.3

1. 22.914
2. 3.2507
3. $8 689.99
4. 7.891
5. 2.037 m
6. 1.1494

Exercise 3.1

1. 200 km
2. $30
3. 16 days
4. 9
5. 64 litres
6. 6 months

Exercise 3.2

1. $8\frac{1}{8}$ kg
2. 32; 28
3. $\frac{1}{18}$
4. $90
5. $160; $64
6. 27; 15

Exercise 3.3

1. $55.13
2. 4.56 m; 13.68 m
3. 77 cents
4. 44.19 kg
5. 60 litres; $72
6. 12

Exercise 3.4

1. (a) $2^3 \times 3^2$ (b) $2^2 \times 3^3$ (c) $2^2 \times 5^2$ (d) $2^2 \times 7^2$ (e) $2^3 \times 3^3$ (f) $2^2 \times 11^2$
2. (a) 2^7 (b) 3^8 (c) 5^5 (d) $2^8 \times 3^4$ (e) $5^3 \times 7^{-4}$ (f) 5

3 (a) 2^3 (b) 3^4 (c) 5^2 (d) 2 (e) $5^2 \times 7^{-2}$ (f) 1

4. (a) 64 (b) 3 (c) 2 (d) 125 (e) 7 (f) 3
5. (a) 128 (b) 45 (c) 16 (d) 500 (e) 2 (f) 1
6. (a) 125 (b) 64 (c) 27 (d) 10 000 (e) 12 (f) 63

Exercise 3.5

1. 135 km
2. $1 890
3. $64.43
4. $972
5. 8 min.
6. $800

Exercise 3.6

1. $424
2. 10 cm; 7.5 cm
3. 0.3 km
4. $33
5. 40
6. 3:4

Exercise 3.7

1. $200
2. 9 litres
3. $3\frac{1}{2}$ min.
4. 9 kg
5. 20 runs
6. 15 min.

Exercise 3.8

1. (a) X = 4, 7 Y = 10, 12
   (b) 2
   (c) Y = 2X
   (d)

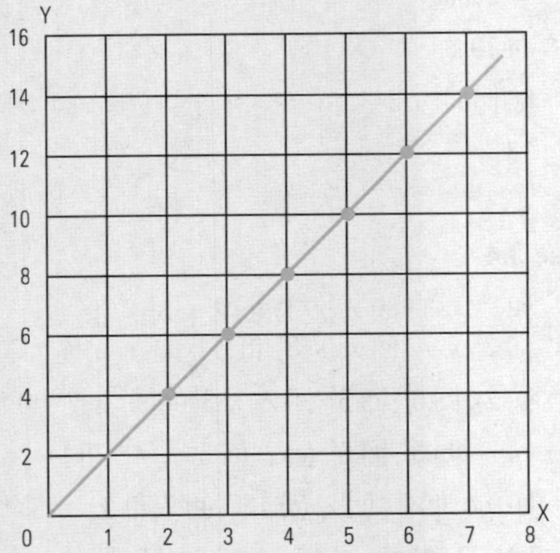

2. (a) X = 12 Y = 18, 45
   (b) 3
   (c) Y = 3X
   (d)

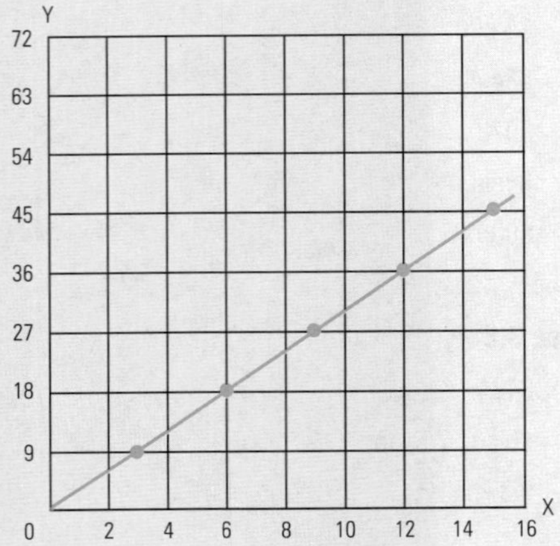

3. (a) X = 3, 5 Y = 1, 2
   (b) ½
   (c) Y = 2X
   (d)

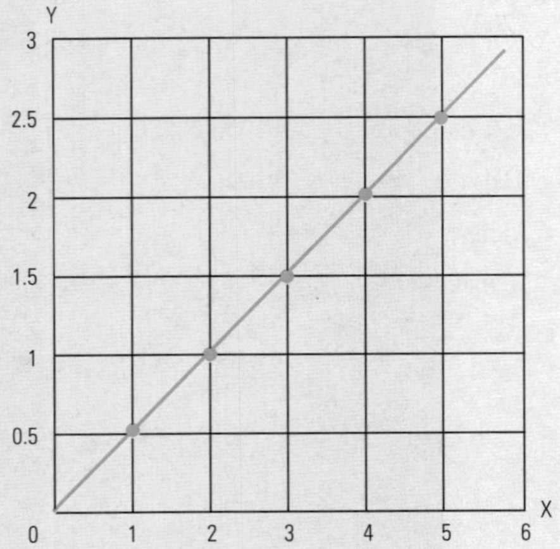

4. (a) X = 5 Y = -9, -10.5
   (b) -1.5
   (c) Y = -1.5X
   (d)

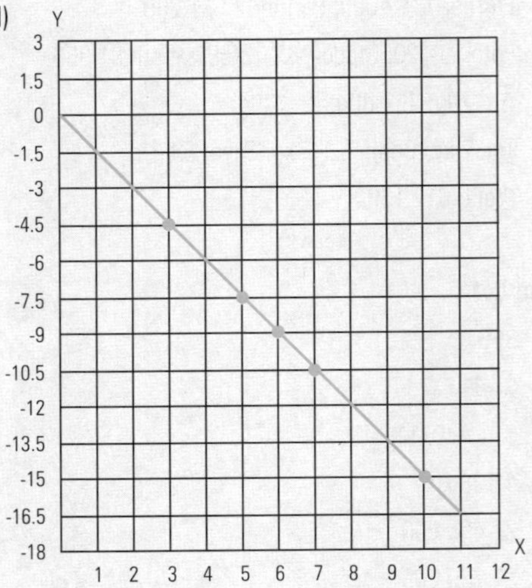

5. (a) 5 (b) $a = 5b^2$ (c) 5, 45, 125, 245, 405 (d) a = 31.25
6. (a) 1.5 (b) $m = 1.5\,n^3$ (c) 1.5, 12, 40.5, 187.5, 514.5

Exercise 3.9

1. ✓
2. ✘
3. ✓
4. ✘
5. ✓
6. ✘

Exercise 3.10

1. 6.5%
2. 1min. 40 sec.; 18 litres
3. b = 5
4. 120 m
5. (a) 68 kg; 180 cm.
6. 1.25 m

2. **Place a tick [✓] in the box if the statement is true, otherwise place a cross [✗] in the box.**

a)	2 : 3 :: 4 : 9	☐	d)	15 : 25 :: 5 : 9	☐
b)	7 : 10 :: 21 : 30	☐	e)	24 : 36 :: 2 : 3	☐
c)	9 : 19 :: 18 : 38	☐	f)	7.5 : 22.5 :: 1 : 3	☐

3. **Place a tick [✓] in the box if a pair of ratios is in direct proportion, otherwise place a cross [✗] in the box.**

a)	$\frac{1}{4} = \frac{3}{12}$	☐	d)	$\frac{5}{6} = \frac{10}{12}$	☐
b)	$\frac{2}{9} = \frac{4}{10}$	☐	e)	$\frac{2}{5} = \frac{8}{9}$	☐
c)	$\frac{3}{8} = \frac{4}{11}$	☐	f)	$\frac{3}{4} = \frac{18}{24}$	☐

4. **Find the value of the unknown in the following direct proportions.**

a)	7 : 9 :: ? : 27	☐	d)	20.5 : 41 :: ? : 2	☐
b)	13 : 5 :: 39 : ?	☐	e)	17 : 5 :: 34 : ?	☐
c)	3.5 : 7 :: 1 : ?	☐	f)	10.8 : 9 :: ? : 1	☐

5. **When a car is filled with 40 litres of petrol it costs $48. How many litres would be filled for $60? (Assume that cost increases in direct proportion to volume of petrol.) The answer is obtained by formulating as follows:**

40 litres : $48 :: ? : $60

Therefore, Number of litres for $60 = ☐ **=** ☐

(Hint: $48 becomes $60 when we multiply $48 by $\frac{60}{48}$.)

6. **What sum of money when increased by 20% becomes $7.20? The answer is obtained by formulating as follows:**

100 : ? = 120 : $7.20

Therefore, the sum of money before 20% increase = ______

= ______

(Hint: To get 100 from 120, we multiply 120 by $\frac{100}{120}$.)

Proportional Division

Dividing an amount in an agreed ratio is called proportional division.

Example (i): Divide $50 between Jack and Jill in the ratio 2: 3

The ratio 2:3 tells us that Jack gets 2 parts and Jill gets 3 parts and there are 5 parts altogether.

5 parts = $50. Therefore, 1 part is $50 5 = $10.

Jack gets 2 parts = 2 x $10 = $20

Jill gets 3 parts = 3 x $10 = $30

Total = $50

Example (ii): Divide $108 between John, Mary and Susan in the ratio 4: 3: 2

The ratio tells us that John gets 4 parts, Mary gets 3 parts and Susan gets 2 parts and there are altogether 9 parts.

9 parts = $108. Therefore, 1 part is $108 9 = $12

John gets 4 parts = 4 x 12 = $48

Mary gets 3 parts = 3 x 12 = $36

Susan gets 2 parts = 2 x $12 = $24

Total = $108

Example (iii): A certain amount of money was shared between Andrew, Betty and Chris in the ratio 5:3:2. If Chris received $800, how much money was shared?

How much did Andrew and Betty receive?

The ratio tells us that Andrew received 5 parts, Betty 3 parts and Chris 2 parts and there are altogether 10 parts.

Chris received 2 parts which represents $800.

Therefore 1 part is $800 / 2 = $400

Amount of money shared = 10 parts = 400 x 10 = $4 000

Andrew received 5 parts = 400 x 5 = $2 000

Betty received 3 parts = 400 x 3 = $1 200

(Check that all three received $4 000 in total.)

Exercise 1.12 (Proportional division)

1. Divide $175 between Abel and Arvo in the ratio 3:4.

7 parts = $ ☐ **. Therefore, 1 part =** ☐

Abel gets 3 parts = ☐ **= $** ☐

Arvo gets 4 parts = ☐ **= $** ☐

2. Share $3 200 between two partners in the ratio 5:3.

8 parts = $ ☐ **. Therefore, 1 part =** ☐

Partner 1 gets 5 parts = ☐ **= $** ☐

Partner 2 gets 3 parts = ☐ **= $** ☐

3. **Divide $7 800 between Peta, Stacey and Elizabeth in the ratio 8 : 7 : 5.**

20 parts = $ ☐ . Therefore, 1 part = ☐

Peta gets 8 parts = ☐ = $ ☐

Stacey gets 7 parts = ☐ = $ ☐

Elizabeth gets 5 parts = ☐ = $ ☐

4. **Three main ingredients A, B and C whose masses are 300 g,100 g and 200 g respectively are to be mixed together to make a product whose mass is 700 g. What should be the individual masses of these ingredients if the mass of the product is to be 980 g?**

Of the 700 g, A's proportion = $\frac{300}{700}$ = ☐

B's proportion = $\frac{200}{700}$ = ☐

C's proportion = $\frac{100}{700}$ = ☐

Mass of A in 980 g = $\frac{3}{7} \times 980$ = ☐

Mass of B in 980 g = ☐ = ☐

Mass of C in 980 g = ☐ = ☐

5. A certain alloy is formed of three metals tin, copper and silver in the ratio 4:3:10. If the mass of the alloy is 8.5 kg, find the mass of each metal in the alloy.

17 Parts = 8.5 kg

Therefore, 1 part = ______ = ______ kg

Tin = 4 parts = ______ = ______ kg

Copper = 3 parts = ______ = ______ kg

Silver = 10 parts = ______ = ______ kg

6. Peter, Mark and Paul are partners in a business sharing profits and losses in the ratio 5 : 2 : 2. If Paul receives $2 700 as his share of the profit this year, what is the total profit of the business for this year?

Paul receives 2 parts which represents $2 700.

Therefore, 1 part is $2 700 ÷ 2 = $ ______

Amount of money shared = 9 parts = ______ = $ ______

Total profit for this year = $ ______

2 Addition and Subtraction

Core learning Outcomes

- Identify and solve addition and subtraction problems involving positive rational numbers.

Exercise 2.1 (Whole numbers)

1. The total of ten numbers is 3 768. The total of first six numbers is 2 212. The total of last five numbers is 1 922. What is the sixth number?

Total of ten numbers =

Total of first six numbers =

Total of last four numbers =

Total of last five numbers =

The sixth number = =

2. The sale in a book store for 31 days is $26 660. If the sale for the first 15 days is $12 788 and the sale for the last 15 days is $12 761, find the sale on the 16th day.

Sale for 31 days =

Sale for the first 15 days =

Sale for the last 16 days =

Sale on the 16th day =

3. **The total wages of 38 employees for a particular week is $22 880. If the total wages of the first 20 employees in the payroll is $76 more than the total income of the last 18 employees, find the total wages of the first 20 employees and that of the last 18 employees in the payroll.**

Total wages of 38 employees = ______

Subtract $76 = ______

Half the difference = ______

Total wages of last 18 employees = ______

Total wages of the first 20 employees = ______

4. **A warehouse received four shipments of merchandise in a month totalling 15 738 kg. The first shipment was 3 216 kg. The second shipment was 1 702 kg more than the first shipment. The third shipment was 1 442 kg less than the second shipment. What was the weight of the fourth shipment?**

Total weight of four shipments = ______

Weight of first shipment = ______

Weight of second shipment = ______ = ______

Weight of third shipment = ______ = ______

Total weight of the first three shipments = ______

Weight of the fourth shipment = ______

5. **The total population of three countries is 50 045 000. If the population of the first country is 15 890 000 and that of the second country is 3 170 000 less than the first country, what is the population of the third country?**

Total population of three countries =

First country =

Second country =

Population of first two countries =

Population of the third country =

6. **A real estate agent sold four houses in a month. The sale value of the first house was $380 000, that of the second house $60 000 more than the first, that of third house $30 000 less than the second house and the sale value of the fourth house is $120 000 more than the first house. What was the total sale value of all four houses?**

First house = $

Second house = $ = $

Third house = $ = $

Fourth house = $ = $

Total sale value of all four houses = $

Exercise 2.2 (Common fractions)

1. Find the sum of $\frac{1}{6}$, $\frac{3}{4}$ and $\frac{5}{8}$.

[ ] = [ ]

2. Add $\frac{3}{7}$ and $\frac{4}{9}$ together and take away the difference between $\frac{4}{9}$ and $\frac{3}{7}$.

$\frac{3}{7} + \frac{4}{9}$ = [ ] = [ ]

$\frac{4}{9} - \frac{3}{7}$ = [ ] = [ ]

Required difference = [ ] = [ ]

3. During a 5 day week, Peter worked the following hours. Monday $6\frac{1}{4}$ hours, Tuesday $7\frac{1}{2}$ hours, Wednesday $5\frac{1}{4}$ hours, Thursday $6\frac{1}{12}$ hours and Friday $7\frac{3}{5}$ hours. How many hours in total did Peter work?

Total hours Peter worked = [ ] = [ ]

= [ ]

4. From the fractions given below, choose the largest fraction and the smallest fraction and add them together.

$1\frac{7}{8}$ $\frac{11}{6}$ $1\frac{5}{9}$ $1\frac{1}{2}$ $1\frac{4}{5}$ = [ ]

(Converting to decimal fractions)

Largest fraction = [ ]

Smallest fraction = [ ]

Total = [ ]

5. The mass of water in a large tank when it is full is to be determined. If the mass of the tank and water is $2\frac{7}{8}$ kg and the mass of the tank alone is $1\frac{2}{3}$ kg, what is the mass of water?

Mass of tank and water =

Mass of tank =

Mass of water = =

6. I bought $5\frac{1}{2}$ kg of fertilser and shared it with two friends. I gave one friend $1\frac{1}{4}$ kg, and the other friend $2\frac{1}{3}$ kg and kept the rest for myself. How many kilograms of fertilser did I keep for myself?

Total mass of fertilser =

Mass of fertilser given to friends = =

Mass of fertilser kept for myself = =

Exercise 2.3 (Decimal fractions)

1. Find the sum of 3.60531, 17.22819 and 2.08092 correct to three decimal places.

=

2. Write as a decimal: $3 + \frac{2}{10} + \frac{5}{100} + \frac{7}{10\,000}$.

=

3. **I had $8 234.60 in my bank account at the end of last month. This month I deposited $1 239.30 into the account and made three online payments, one for telephone $134.56, one for electricity $210.45 and another one for property rates $438.90. What will be the balance in my account now?**

Amount in the bank at the end of last month = $ ______

Add: **Amount deposited this month** = $ ______

Total = $ ______

Subtract: **Payments for telephone** = $ ______

Payments for electricity = $ ______

Payments for rates = $ ______ = $ ______

Balance in the account now = $ ______

4. **Find in decimal form the sum of the following:**

17/20, 11/16, 3.625, 2.725, 0.0035

______ = ______

5. **A piece of string 4.65 m long is to be cut into three pieces of different lengths. If one piece is to be 1.365 m and another 1.248 m, what is the length of the third piece?**

Length of string = ______

Total length of two pieces = ______

Length of the third piece = ______

6. **Find the sum of the decimal fractions 2.1847, 3.1012 and 1.006 and take it away from the sum of 1.2178, 2.1345 and 4.089.**

Sum of ______ = ______

Sum of ______ = ______

Difference = ______

3 Multiplication and Division

Core learning Outcomes

- Identify and solve multiplication and division problems involving positive rational numbers
- Identify and solve multiplication and division problems involving indices
- Identify and solve multiplication and division problems involving rates
- Identify and solve multiplication and division problems involving ratios
- Identify and solve multiplication and division problems involving direct proportions

Exercise 3.1 (Rational numbers)

1. **A train travels 120 km in 3 hours at a steady speed. How far will it travel in 5 hours at the same speed?**

Distance traveled in 1 hour = ______ = ______

Distance traveled in 5 hours = ______ = ______

2. I bought 6 mangoes at a total cost of $12. How much will 15 mangoes cost at the same unit price?

Cost of 1 mango = ☐

Cost of 15 mangoes = ☐

3. Steve reads 25 pages of a story book every day. If there are 400 pages in the book, how long will he take to read the whole book?

No. of days to read 400 pages = =

4. Christine has $75 made up of $5 notes and $10 notes. If she has three $10 notes, how many $5 notes does she have?

Three $10 notes = $ ☐

Value of $5 notes = $ ☐

Number of $5 notes = ☐

5. A motor vehicle uses 8 litres of petrol to travel 72 km. How many litres of petrol would be needed to travel 576 km?

Distance traveled with 1 litre = ☐ = ☐

No. of litres needed to travel 576 km = ☐ = ☐

6. Sam saves $12 every month from the monthly pocket money he gets from his parents to buy a toy he likes. How many months will it take Sam to save up $72 to buy that toy?

No. of months to save $72 = ☐ = ☐

Exercise 3.2 (Common fractions)

1. A parcel weighs $3\frac{1}{4}$ kg and a second parcel weighs $2\frac{1}{2}$ times as much. What is the weight of the second parcel?

   Weight of second parcel = ☐ = ☐

2. If $\frac{3}{4}$ of a number is 24, what is the number? What is $\frac{7}{8}$ of the number?

   Number = ☐ = ☐

   $\frac{7}{8}$ of the number = ☐

3. I saved $\frac{1}{6}$ of my earnings and then spent $\frac{2}{3}$ of it. What fraction of the earnings is still left?

   Savings as a fraction of earnings = ☐

   Fraction spent out of savings = ☐ = ☐

   Fraction of earnings still left = ☐ = ☐

4. Tim and Tammy wanted to share a sum of money. Tim is to receive $\frac{1}{3}$ and Tammy $\frac{2}{3}$. If Tammy gives $\frac{1}{3}$ of her share to the sister Sissy and has $40 left, how much was shared between Tim and Tammy?

   Tammy's share = ☐

   Fraction given to Sissy = ☐

   Fraction left = ☐

   Amount shared between Tim and Tammy = ☐ = ☐

5. Josh went shopping with a certain sum of money. He spent $\frac{2}{5}$ of the money on clothes and $\frac{1}{6}$ of the remainder on books and still had $80 left. How much money did he have at first? How much did he spend on clothes?

Fraction spent on clothes =

Fraction spent on books = =

Total fraction spent =

Fraction still left =

Amount Josh had at first = =

Amount spent on clothes = =

6. In a class $\frac{5}{9}$ of the total students are girls. If there are 12 boys in the class, how many students are in the class? How many are girls?

Girls as a fraction of total =

Boys as a fraction of total =

No. of students in the class = =

No. of girls in the class = =

Exercise 3.3 (Decimal fractions)

1. A typist charges $15.75 per hour of typing. How much will the charge be for 3.5 hours?

Charge = =

2. A string 36.48 metres long is cut into 8 equal pieces. What is the length of each piece? What is the length of 3 such pieces?

Length of each piece = =

Length of 3 pieces = =

3. A printer estimates the cost to print 500 flyers to be $385.60. What is the cost per flyer?

Cost per flyer = =

4. What is the total weight of, 3 parcels weighing 3.24 kg each, 5 parcels weighing 2.75 kg each and 7 parcels weighing 2.96 kg each?

Weight of 3 parcels = =

Weight of 5 parcels = =

Weight of 7 parcels = =

Total weight = =

5. A car can travel 9.25 km on I litre of petrol. How many litres would be needed for traveling 555 km? If the cost per litre of petrol is $1.20, what is the total cost of the petrol?

Litres needed to travel 555 km =

=

Total cost of petrol = = $

6. **The chocolates in a box weigh 150.96 g. If each chocolate weighs 12.58 g. How many chocolates are there in the box?**

No. of chocolates in the box =

=

Rules of Indices

1. When multiplying numbers with powers where the base number is the same, add the powers.

Examples: $2^3 \times 2^4 = 2^7$

$5^5 \times 5^{-3} = 5^2$

2. When dividing numbers with powers where the base number is the same, subtract the powers.

Examples: $2^5 \div 2^3 = 2^{5-3} = 2^2 = 4$

$4^2 \div 4^3 = 4^{2-3} = 4^{-1} = \frac{1}{4}$

3. When a number raised to a power is further raised, multiply the powers.

Examples: $(2^3)^2 = 2^{3\times2} = 2^6$

$(3^2)^{-2} = 3^{-2\times2} = 3^{-4} = \frac{1}{3^4} = \frac{1}{81}$

$(64)^{1/2} = (2^6)^{½} = 2^3 = 8$

When any number is raised to the power zero, its value is 1.

Example:

$3^0 = 3^{1-1} = \frac{3}{3} = 1$

The 'first power' of a number is always equal to itself.

Examples:

$2^1 = 2 \quad 6^1 = 6 \quad 10^1 = 10$

Exercise 3.4 (Indices)

1. Write the following as product of two or more base numbers each raised to a power.

Example: $36 = 2 \times 2 \times 3 \times 3 = 2^2 \times 3^2$

a) **72** ______ = ____

b) **108** ______ = ____

c) **100** ______ = ____

d) **196** ______ = ____

e) **216** ______ = ____

f) **484** ______ = ____

2. Use the multiplication rule of indices to simplify the following (leave the answer in index form):

a) $2^4 \times 2^3$ = ______

b) $3^3 \times 3^5$ = ______

c) $5^{-2} \times 5^7$ = ______

d) $2^5 \times 3^5 \times 2^3 \times 3^{-1}$ = ______

e) $7^{-3} \times 5^0 \times 7^{-1} \times 5^3$ = ______

f) $11^2 \times 5^5 \times 5^{-4} \times 11^{-2}$ = ______

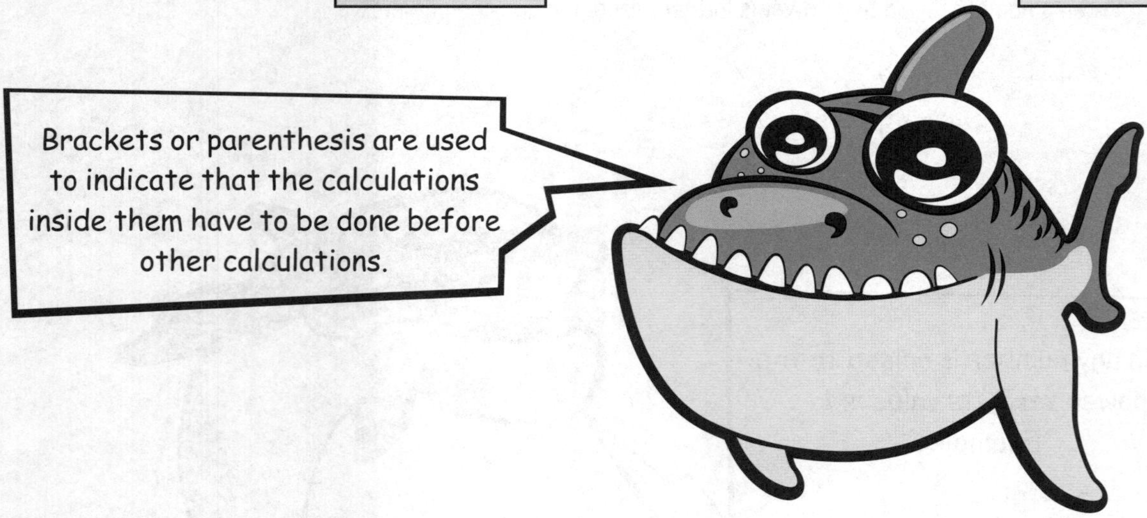

3. Use the division rule of indices to simplify the following (leave the answer in index form):

a) $2^5 \div 2^2$ = ______ = ____

b) $3^2 \div 3^{-2}$ = ______ = ____

c) $5^{-4} \div 5^{-6}$ = ______ = ____

d) $(2^6 \div 2^3) \div (2^3 \div 2)$ = ______ = ____

e) $(5^{-2} \div 5^{-4}) \div (7^{-1} \div 7^{-3})$ = ______ = ____

f) $(3^8 \div 3^{-2}) \div 3^6$ = ______ = ____

4. Apply the rule of indices for raising to a power a number already raised to a power. Leave the answer in its simplest form.

a) $(2^3)^2$ = ☐ = ☐

b) $(27)^{1/3}$ = ☐ = ☐

c) $(16)^{1/4}$ = ☐ = ☐

d) $(25^{1/2})^3$ = ☐ = ☐

e) $(343)^{1/3}$ = ☐ = ☐

f) $(243)^{1/5}$ = ☐ = ☐

5. Apply the rules of indices and simplify the following:

a) $2^4 \times 2^3$ = ☐ = ☐

b) $5^3 \times 3^4 \div 15^2$ = ☐ = ☐

c) $(2^2)^5 \div (2^3)^2$ = ☐ = ☐

d) $(25^{1/2})^3 \times 4$ = ☐ = ☐

e) $(6^3 \div 2^2) \times 3^{-3}$ = ☐ = ☐

f) $8^{-2} \times 4^3 \times 7^0$ = ☐ = ☐

6. **Without using a calculator simplify the following:**

a) $\frac{5^2 \times 5^4}{5^3}$ = ______ = ______

b) $\frac{(4 \times 2^2 \times 3)^2}{36}$ = ______ = ______

c) $\frac{3^2 \times 3^4 \times 3^6}{3 \times 3^3 \times 3^5}$ = ______ = ______

d) $\frac{10 \times 10^2 \times 10^4}{10^3}$ = ______ = ______

e) $\frac{2^5 \times 3^2 \times 9}{2^2 \times 3^3 \times 2}$ = ______ = ______

f) $\frac{(49 \times 3^3)^2}{343 \times 3^4}$ = ______ = ______

Exercise 3.5 (Rates)

1. **A car traveled at a uniform speed of 90 km per hour for 1½ hours non-stop. What distance did it travel?**

______ = ______ km

2. **An employee is paid $42 per hour for working on a certain project. If the total hours worked is 45 hours, what will be the total pay?**

______ = $ ______

3. **If petrol costs $1.18 per litre, how much will it cost to buy 54.6 litres?**

______ = $ ______

4. **If tax is deducted from my earnings at the rate of 30 cents in the dollar, how much tax would be deducted if my earnings are $3 240?**

______ = $ ______

5. **Water is flowing from a tap at the rate of 110 ml per second. How many minutes will it take to fill a container with a capacity of 52.8 litres?**

[ ] = [ ] **minutes**

6. **A bank pays customers interest at the rate of 5% per year on amounts kept in a deposit account. How much interest will be paid if $8 000 is kept in the account for 2 years?**

[ ] **= $** [ ]

Increasing an amount, quantity or size in a certain ratio, say 4:3, means the new value becomes 4/3 times that of the original.

Decreasing an amount, quantity or size in a certain ratio, say 1:5, means the new value becomes 1/5 times that of the original.

Exercise 3.6 (Ratios)

1. **An employer decided to increase all the employees' wages in the ratio 4:3. If John earns $318 per week at present, how much will he earn per week after the increase?**

Wages per week at present **= $** [ ]

Wages per week after increase **=** [ ] **= $** [ ]

2. **A photograph 8 cm long and 6 cm wide is enlarged in the ratio 5:4. What are its new length and width?**

New length **=** [ ] **=** [ ] **cm**

New width **=** [ ] **=** [ ] **cm**

3. **A map has been drawn to a scale of 1:15 000. The distance between two schools in a suburb is shown in the map as 2 cm. What is the true distance between the schools in kilometres?**

True distance between schools = [ ] = [ ] **km**

4. **The sale price of a shirt was reduced in the ratio 6:7. If the old price was $38.50, what is its reduced price?**

Reduced price of shirt = [ ] = $ [ ]

5. **In a class the ratio of girls to boys is 3:2. If there are 8 more girls than boys, how many students are in the class?**

Proportion of girls = [ ]

Proportion of boys = [ ]

Difference in Proportion = [ ]

Total number of students = [ ] = [ ]

6. **In a camp, 180 soldiers have food for 20 days (i.e 1 soldier can survive for 3 600 days). After eight days, 60 more soldiers joined the camp. In what ratio should the rations be reduced per head per day to make the remaining food last for 12 days?**

No. of days rations available for 1 soldier after 8 days = [ ] = [ ]

No. of days rations required for 1 soldier now = [ ] = [ ]

Required ratio reduction for rations = [ ] = [ ]

Direct Proportions

Problems involving direct proportion are easily solved by finding a value for one item (or one unit) first by dividing and then multiplying to find the value of many items (or units). This method is sometimes called the unitary method.

Example: If 6 compact disks cost $18, what will be the cost of 15 disks?

6 compact disks = $18

1 compact disk = 18/6 = $3

15 compact disks = 15 x $3 = $45

Exercise 3.7 (Direct Proportions)

1. The cost of 8 digital watches is $320. What is the cost of 5 such watches?

8 digital watches = $ ☐

1 digital watch = $ ☐ = $ ☐

5 digital watches = $ ☐ = $ ☐

2. Water is poured into a long cylindrical container. When the height of water is 18 cm, the volume is 6 litres. What will be the volume when the height is 27 cm?

When height is 18 cm, volume = ☐

When height is 1 cm, volume = ☐ = ☐ **litres**

When height is 27 cm, volume = ☐ = ☐ **litres**

3. **A photocopier makes 80 copies in 2 ½ minutes. How long will it take to make 112 copies?**

Time taken for 80 copies =

Time taken for 1 copy = =

Time taken for 112 copies = =

4. **In a shop, 3 kg of grapes cost $12.60. How many kilograms of grapes can be bought for $37.80?**

$12.60 = **3 kg**

$1 = =

$37.80 = =

5. **A batsman in a cricket game scores 5 runs every 2 overs. If he continues at this rate, how many runs can he score in 8 overs?**

In 2 overs runs scored =

In 1 over average runs scored =

In 8 overs runs scored = =

6. **A typist can type 496 words in 8 minutes. How long will it take to type 930 words at the same speed?**

Time taken to type 496 words =

Time taken for 1 word = =

Time taken for 930 words = =

Graphs and Direct Proportions

When two quantities are in direct proportion we have a constant ratio between each pair of values of the two quantities.

Example (i): The grid below shows the relationship between a quantity X and another quantity Y. X is an independent variable and Y a dependent variable, as it depends on the value X takes.

X	2	4	6	8	10
Y	6	12	18	24	30

We observe that each value of Y is 3 times the corresponding value of X.

That is, there is a constant ratio between the values of Y and the corresponding values of X.

The quantity Y is said to be in direct proportion to X because of this relationship and 3 is called the constant of proportion.

The statement, Y is proportional to X is usually written as:

$Y \propto X$

Or $Y = k \times X$ (k stands for the constant of proportion)

When say, $X = 2$, $Y = 6$; i.e $6 = k \times 2$

$k = \frac{6}{2} = 3$

The relationship between Y and X is written as: $Y = 3X$

If the values of Y are plotted against the values of X, a straight line passing through the origin will be obtained as shown below.

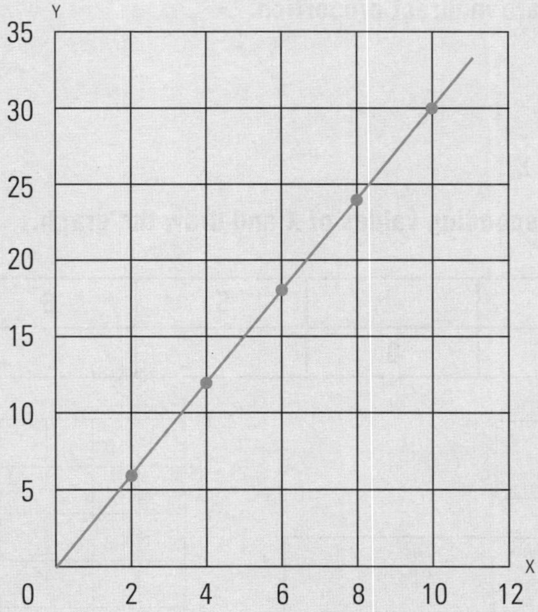

Y can also vary directly as X^2, or directly as X^3, or directly as $\sqrt{X}$ and so on.

Example (ii): Y varies directly as X^2.

X	2	4	6	8	10
X^2	4	16	36	64	100
Y	8	32	72	128	200

We observe that each value of Y is 2 times the corresponding value of X^2 (and not X).

Here, 2 is called the constant of variation. It is obtained by dividing Y by X^2.

Example (iii): Y varies directly as X^3.

$Y \propto X^3$ (or $Y = kX^3$)

X	2	4	6	8	10
X^3	8	64	216	512	1 000
Y	8	64	216	512	1 000

We observe that each value of Y is equal to the corresponding value of X^3 (and not X).

Here, the constant of variation is obtained by dividing Y by X^3 and is equal to 1.

Exercise 3.8 (Direct Proportions)

In each of the following four cases X and Y are in direct proportion.

a) Insert the missing X and Y values.

b) Calculate the constant of proportion.

c) State the relationship between Y and X.

d) Plot the values of Y against the corresponding values of X and draw the graph.

1.

X	2	3		5	6	
Y	4	6	8			14

Constant of proportion =

Relationship between Y and X is:

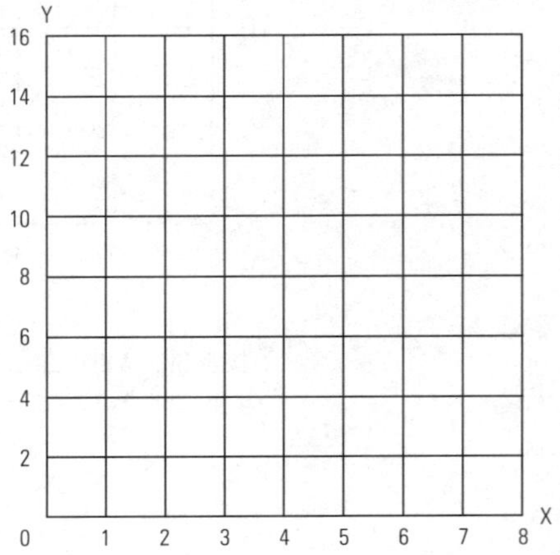

2.

X	3	6	9		15
Y	9		27	36	

Constant of proportion =

Relationship between Y and X is:

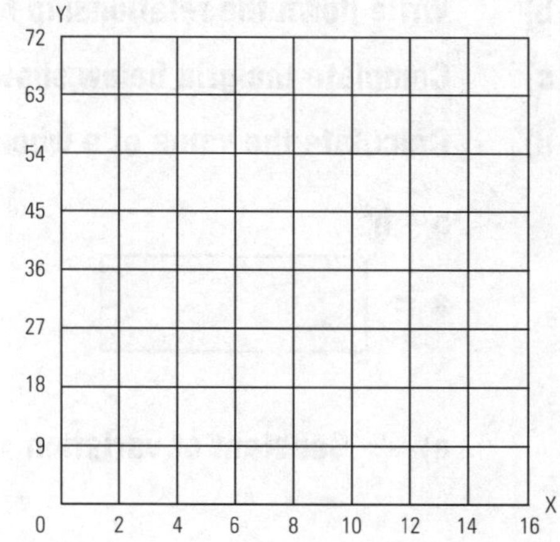

3.

X	1	2		4	
Y	0.5		1.5		2.5

Constant of proportion =

Relationship between Y and X is:

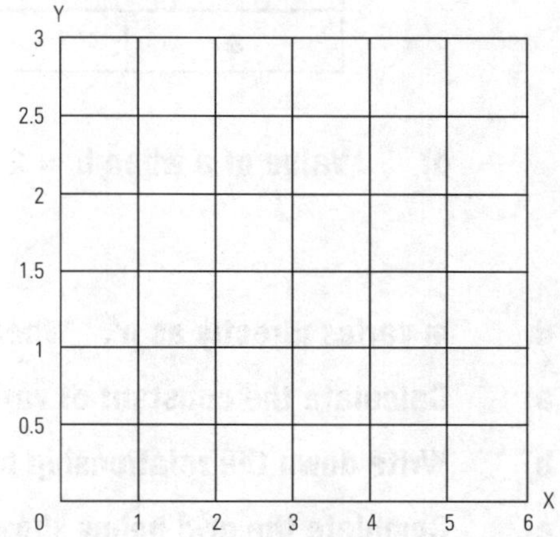

4.

X	3		6	7	10
Y	-4.5	-7.5			-15

Constant of proportion =

Relationship between Y and X is:

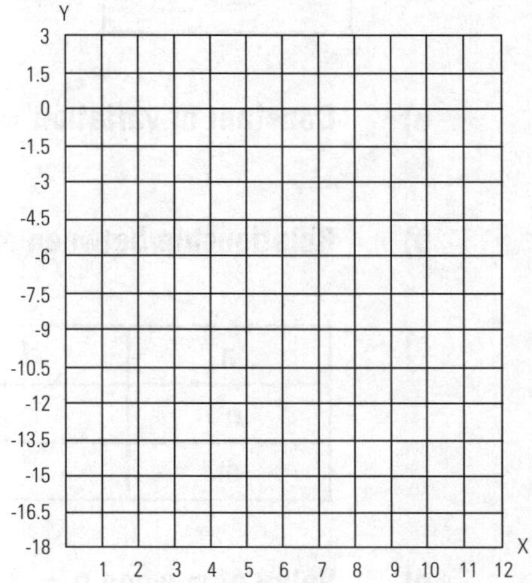

5. Two quantities a and b are in proportion such that: $a \propto b^2$.

a) Calculate the constant of variation if a = 80 when b = 4.

b) Write down the relationship between a and b.

c) Complete the grid below showing the values of a for the given values of b.

d) Calculate the value of a when b = 2.5

$a \propto b^2$

a =

a) Constant of variation =

b) Relationship between a and b is:

c)

b	1	3	5	7	9
b^2					
a					

d) Value of a when b = 2.5 is:

6. m varies directly as n^3. When n = 2, m = 12.

a) Calculate the constant of variation

b) Write down the relationship between m and n

c) Complete the grid below showing the values of m for the given values of n.

d) Calculate the value of m when n = 8

$m \propto n^3$

m =

a) Constant of variation =

b) Relationship between m and n is:

c)

n	1	2	3	5	7
n^2					
m					

d) Value of m when n = 8 is:

Exercise 3.9 (Direct Proportions)

For each pair of quantities given, place a tick [✓] in the box if they are in direct proportion. Otherwise place a cross [✗] in the box.

1.

b	1	4	6	8
a	2.5	10	15	20

2.

q	3	6	8	10
p	6	9	16	20

3.

n	-2	3	4	6
m	6	-9	-12	-18

4.

h	4	8	12	16
w	2	4.5	6	18

5.

v	1	1.5	2	4
c	10	15	20	40

6.

s	2	8	18	32
t	1	2	3	4

Exercise 3.10 (Direct proportion/variation)

1. **The simple interest (I) earned by a given sum of money for a fixed term varies as the rate percent per annum (r). If the interest is $80 when the interest rate is 5%, what rate percent is necessary to produce $104 interest?**

   $I = k \times r$

   When $r = 5$, $I = 80$ $80 =$

   Therefore, $k =$

   When $I = 104$, $r = \frac{104}{k} =$

   $=$

2. The time (t) taken by a tap to fill a large vessel is directly proportional to the volume (v) of water. Write down the relationship between t and v. If t = 80 seconds when v = 32 litres, how long will it take to fill the vessel with 40 litres? How many litres would be filled in 45 seconds?

$t = k \times v$

When t = 80, v = 32 — 80 =

Therefore, k =

When v = 40, t =

=

When t = 45, v =

=

3. If a varies as the cube of b and b = 3 when a = 81, what value of b will make a = 375?

$a = k \times b^3$

b = 3 when a = 81 — 81 =

k =

When a = 375, b =

=

4. **The distance (h) travelled by a body falling from rest under gravity is directly proportional to the square of the time for which it falls. If the body falls 43.2 metres in 3 seconds, what distance does it travel in 5 seconds?**

$h = k \times t^2$

h = 43.2, when t = 3   43.2 =

k =

When t = 5,   h =

=

5. **The mass (m kg) of a person approximately varies in direct proportion to the height (h cm). When the height is 162 cm, the mass is 72 kg. Calculate (a) the mass when the height is 153 cm (b) the height when the mass is 80 kg.**

$m = k \times h$

h = 162, when m = 72   72 =

k =   =

When h = 153,   m =

=

When m = 80,   h =

=

6. **The length (L metres) of a simple pendulum varies as the square of its time of a complete swing (t seconds). The complete swing time of a pendulum of length 1.8 m is 3 seconds. Find the length in metres of a pendulum the time of whose swing is 2.5 seconds.**

$L = k \times t^2$

$L = 1.8$, when $t = 3$ 1.8= ______

k = ______

When $t = 2.5$, **L =** ______

= ______

Notes

Notes